THE CLOCK TOWER

A History of Benicia's Mighty Fortress

The Clock Tower

A History of Benicia's Mighty Fortress

H. Allan Gandy

Alive Book Publishing

The Clock Tower
A History of Benicia's Mighty Fortress
Copyright © 2021 by H. Allan Gandy

All rights reserved. No part of this book may be reproduced or transmitted in any form or by any means without written permission from the publisher and author.

Additional copies may be ordered from the publisher for educational, business, promotional or premium use. For information, contact ALIVE Book Publishing at: alivebookpublishing.com, or call (925) 837-7303.

Book Design by Alex P. Johnson

ISBN 13
978-1-63132-151-1

Library of Congress Control Number: 2021919339
Library of Congress Cataloging-in-Publication Data is available upon request.

First Edition

Published in the United States of America by ALIVE Book Publishing and ALIVE Publishing Group, imprints of Advanced Publishing LLC
3200 A Danville Blvd., Suite 204, Alamo, California 94507
alivebookpublishing.com

PRINTED IN THE UNITED STATES OF AMERICA

10 9 8 7 6 5 4 3 2 1

Acknowledgements

I would like to express my special thanks to the Benicia Historical Museum for the opportunity to research this project on the Clock Tower and their permission to use the incredible photos from their archives.

Many thanks to Rick Knight, City of Benicia, for his information and photos on the Seth Thomas clock operating mechanism.

The Clock Tower
A History of Benicia's Mighty Fortress
by H. Allan Gandy

Cover photograph: Clock Tower storehouse, June 1904
Source: Benicia Historical Museum

To Kylie, my wonderful granddaughter

Contents

The Clock Tower 11

New Storehouse for the Benicia Arsenal 13

The Clock Tower Design 15

       The Main Entrance 18

       The Cannon Ports 20

       The Windows 21

       The Walls 22

       The Towers 23

       The Interior 24

The Builders of the Clock Tower 27

The Construction of the Clock Tower 31

The Explosion and Fire of 1922 43

Rebuilding of the Clock Tower 51

The Clock 55

The Tunnel 59

The Clock Tower Today 61

Appendix A – List of men who built the Clock Tower 63

The Clock Tower

Figure 1: The Clock Tower, 2020.
Source: Photograph by author.

Constructed in 1859, this US Army fortified storehouse was originally constructed as a three-story stone building using native Benicia sandstone. It was the first military fortress built in the West. An explosion and fire in 1912 gutted the structure and it was rebuilt to the two-story structure called the Clock Tower that exists today (Figure 1).

The Clock Tower is listed in the United States Department of the Interior, National Register of Historic Places, dated November 7, 1976. It is described as:

Storehouse (Clock Tower) (Building #29)

This sandstone building was constructed in 1859 and was the first military fortress built in the West. The walls are 2' 6" thick and the ceiling height is 14'. Building dimensions are 61' x 177.3' with 18,600 square feet of usable space. The lower floor is concrete and the upper is of wood. The roof is gabled and the roofing is asbestos cement-shingled. Originally, the fortress had three stories with a crenelated roof, two look-out towers, one at the southwestern corner and one at the northwest corner. Two small decorative towers graced the remaining corners. Apertures on the west and east sides of the building were designed for howitzers and the long loophole windows for musketry. The structure was gutted by fire and explosion in 1912 and was restored as a two story building. The front tower was undamaged. A large American-made Seth Thomas clock was installed on it as a memorial to Col. Julian McAllister who commanded the Arsenal from 1860-1864 and 1867-1886. The clock was operated by a heavy metal weight on a cable that slowly unwound on a windlass. A crank was used to wind the cable weekly. It was operative as late as June 14, 1967.

A tunnel from the commander's house connected with this fortress.

During its long history the fortress served as quarters for enlisted men during the Civil War, a storehouse, a munitions depot, chapel, and a National Guard Armory. Location is on a point at the eastern end of Washington Street,[1]

The building was called the "main storehouse" by the Army. It became known as the Clock Tower after re-installation of large clocks in 1937 honoring longtime arsenal commander Colonel Julian McAllister.

New Storehouse for the Benicia Arsenal

When Franklin D. Callender took over command of the Benicia Arsenal in 1856, he requested and received an appropriation of $50,000 to be set aside for a new storehouse. On June 18, 1857, he wrote to Colonel H. K. Craig, ordnance officer in Washington D. C., suggesting that the most pressing need of the arsenal was a new storehouse and urged that the appropriation recently been allotted for the arsenal be used for that purpose. Callender requested drawings of recent storehouses for his study.[2]

On November 16, 1857, he reported that arsenal officer Colonel J. W. Ripley had selected a site. He was concerned with the building's appearance, suggesting that it be executed in the fine local sandstone and wrote that it should be "as handsome a building as is proper for the purpose for which it is designed, having a due regard for economy and the appropriation made for that purpose." Callender then asked for drawings of plans prepared for a stone building at the Arsenal in Watervliet, New York, and the storehouse for small arms at the Springfield Armory. Callender subsequently received these drawings but in a letter of November 18, 1857, stated that they were not exactly what was needed. He wrote a letter to Alfred Mordecai, commander of the Watervliet Arsenal, asking for advice and requesting more drawings. He mentioned that he believed there was an arsenal building in New York City with towers on each corner and asked for a sketch of that building. The reason for the towers, he explained, was to prevent any possibility of attack on the storehouse by local citizens. By this time he obviously had a clear idea of the kind of structure he wished. One of the ideas for the arsenal building was fortification for defense against any sudden attack, since the building for small arms at this arsenal would contain nearly all the arms on the West Coast, and recently during the vigilance committee activities in San Francisco, serious apprehensions were entertained that an attempt would be made to seize them.[3]

The site for the location of the new Arsenal storehouse was selected by Captain Callender and other Army officers. It was to be located atop a hill 80 feet in elevation at Army Point, a strategic location where the Suisun Bay meets the Carquinez Strait (Figure 2). However, this site currently had a structure, the Quartermasters Residence, which was under the control of the Quarter Masters' Department. Callender wrote to Colonel Thomas Swords, Deputy Quartermaster General, in March 1857 asking for this site to be transferred to the Ordnance Department in exchange for other Arsenal "useful grounds" and enclosed a map of the proposed relocation (Figure 3). The request was approved by Jefferson Davis, Secretary of War, in March 1858 and by Colonel H. K Craig, Chief of Ordnance, in March 1859.

Figure 2: A drawing made in 1857 with Quartermaster's Residence shown in the center. This would be the location of the new storehouse. Mount Diablo is in the background.
Source: Benicia Historical Museum.

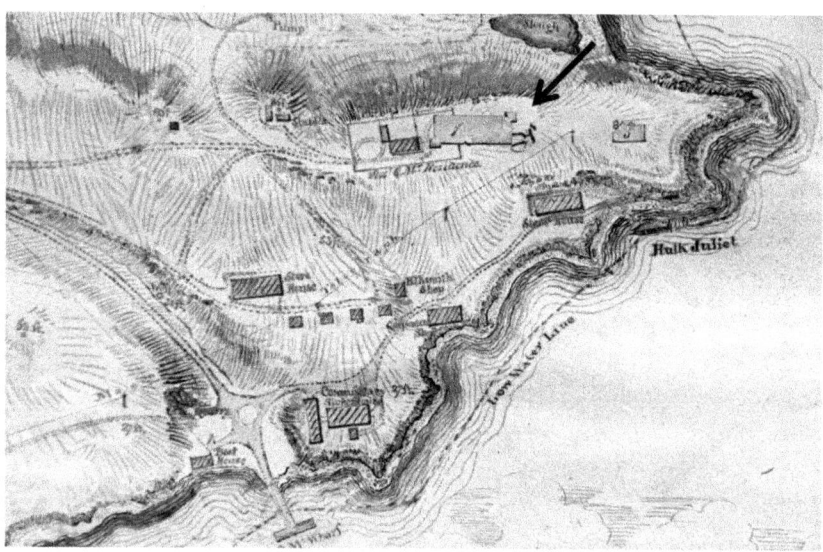

Figure 3: Map of Army Point on a 1857 drawing with the storehouse location drawn over the Quartermaster's Residence (shown with an arrow).
Source: Benicia Historical Museum.

THE CLOCK TOWER DESIGN

Commander Franklin D. Callender's correspondence to the Ordnance Office shows his interest in the aesthetic as well as functional aspects of the new storehouse design. He wanted a building with towers on all four sides to be used for flank defense in the case of attack by local citizens. Two of the four towers were deleted and replaced with turrets at the final stage of the design. This was probably justified from a functional point of view, however it certainly changed the monumental symmetry of the original design.

The storehouse was constructed from local Benicia sandstone that was very carefully cut and joined. The building dimensions are 61 feet by 177 feet with walls 2 feet 6 inches thick. There is 18,600 square feet of usable space in the three-story structure. The foundation is sandstone forming a water table on the south side of the building and around the two towers. A water table is a masonry feature, sometimes decorative, that consists of a projecting layer (course) that deflects water running down the face of a building away from the foundation.

The drawings for the Arsenal storehouse were drawn by master builder Jeremiah Fuss with input from Callender. The drawing in Figures 4 and 5 show the original concept of the building with four towers. The drawing shown in Figure 6 shows the final two-tower configuration.

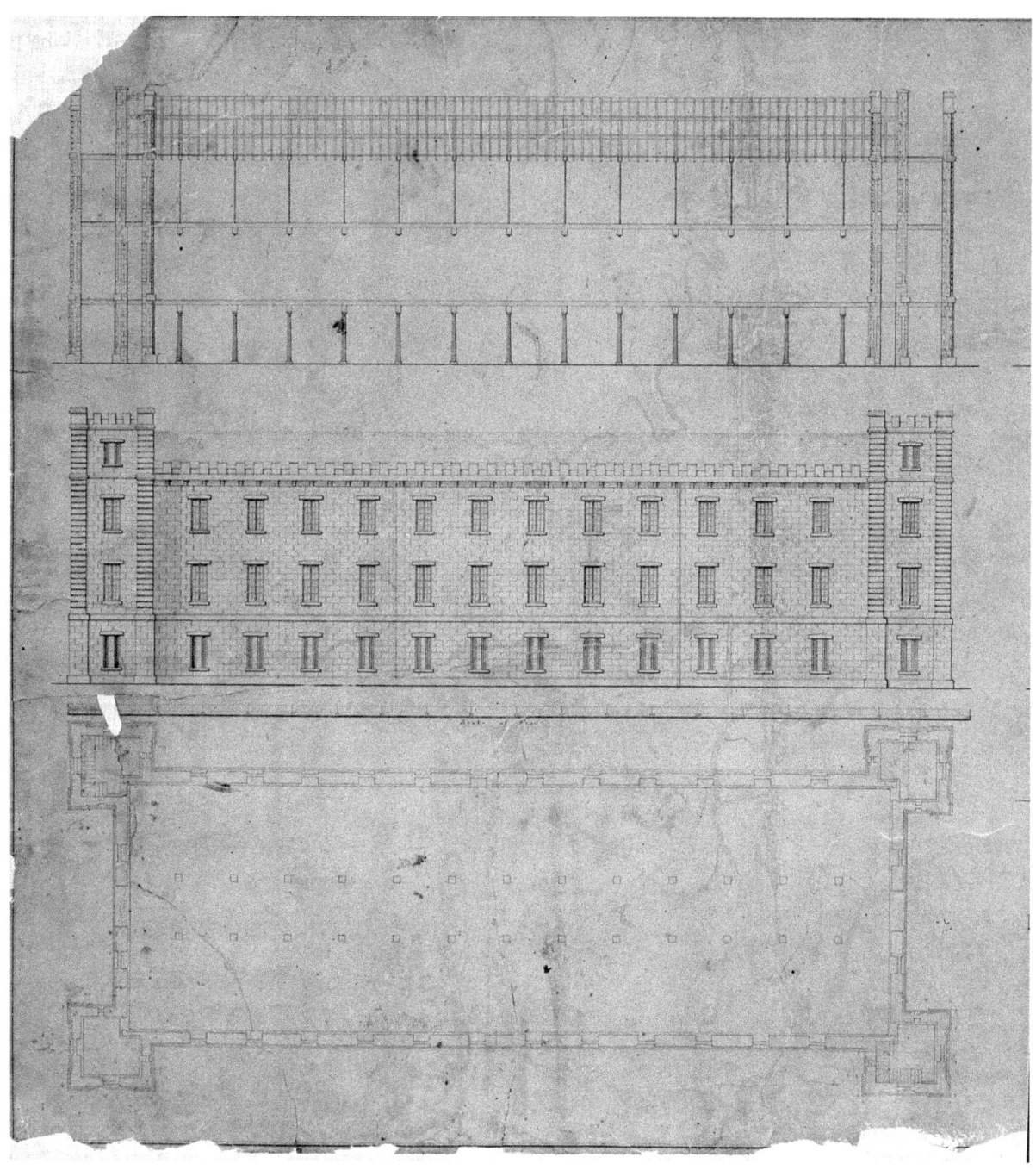

Figure 4: Drawing for the new storehouse, 1859.
Source: Library of Congress, HABS no. CA-1828-13.

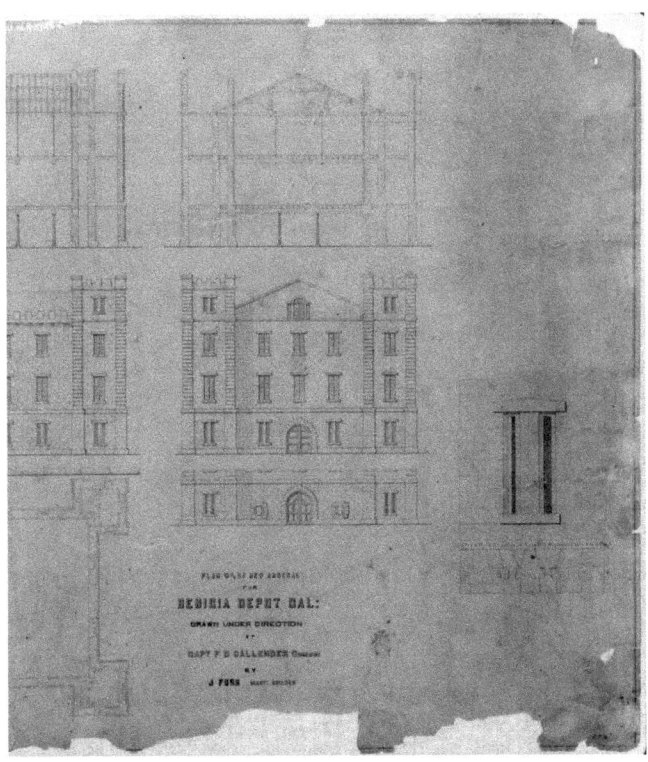

Figure 5: Drawing for the new storehouse, 1859.
Source: Library of Congress, HABS no. CA-1828-12.

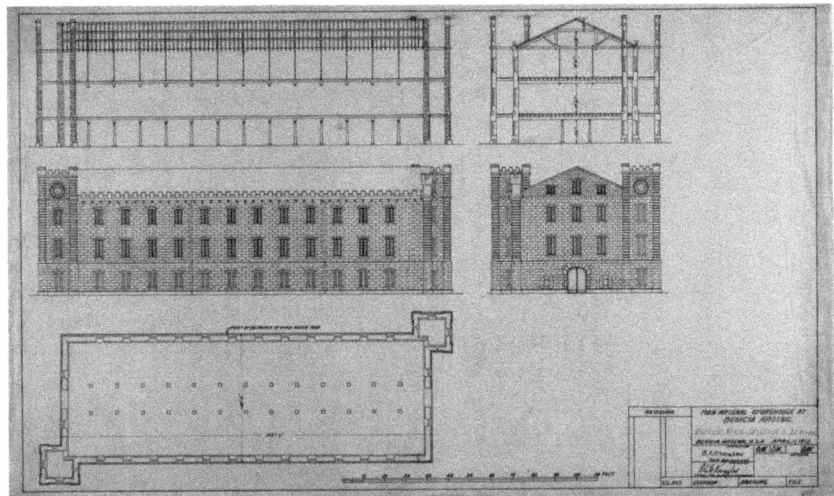

Figure 6: Drawing of the storehouse, April 11, 1912.
Source: Library of Congress, HABS no. CA-1828-17.

The Main Entrance

The main entrance is in the center of the west side wall. An identical entrance is on the east side. The entrance is a large three-centered segmental arch surrounded by quoins. Carved in the sandstone keystone is "Erected AD 1859" (Figure 7).

Figure 7: Keystone "Erected AD 1859."
Source: Photograph by author.

The door is a large arched, steel door (Figure 8) with iron strap hinges and iron slots for a large wooden bolt. In the center of the right-hand leaf on the entrance, a small passage door has been cut for access (Figure 9).[4]

Figure 8: Drawing of the storehouse door, 1859.
Source: Library of Congress, HABS no. CA-1828-15.

Figure 9: Clock Tower main door.
Source: Photograph by author.

Figure 10: Of course you can't open the door without the Clock Tower key.
Source: Benicia Historical Museum.

The Cannon Ports

On both sides of the entrance doors (front and back) are small square openings, called cannon ports (Figure 11 and 12), with decorative stones that has metal closures with locking bolts.

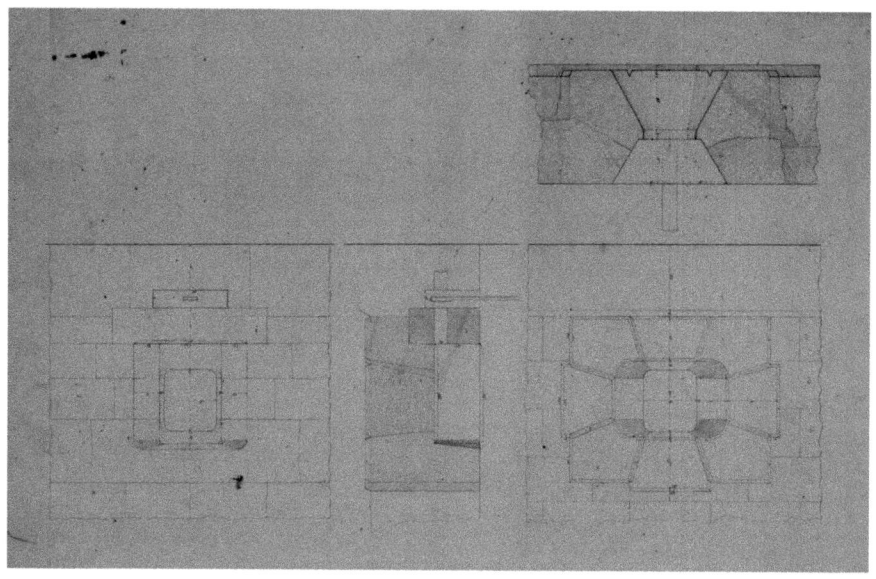

Figure 11: Cannon port drawing, 1859.
Source: Library of Congress, HABS no. CA-1828-16.

Figure 12: Cannon port.
Source: Photograph by author.

The Windows

There are twelve windows on the each side on the first story of the storehouse. The windows are called "loop-hole" windows which have two small vertical slit openings, six inches wide, with a stone mullion between (Figure 13 and 14). A mullion is a vertical stone structure between the window openings. The openings were hinged metal closures on the inside with bolt locks. The decorative stone atop the windows are called lintels and are made from one large piece of stone carved in a pediment shape.

Figure 13: First floor windows.
Source: Photograph by author.

Figure 14: Loop-hole window with lintel above.
Source: Photograph by author.

The Walls

The main walls of the building are rusticated sandstone with water table, belt course at second-floor line and quoins (decorative end stones) at the corners. The quoins are rock-faced on the first floor and smooth-faced above (Figure 15). The face of the stones are "rock-faced" rough as they came from the quarry, and with a smooth margin (Figure 16). This rough finished provided a more defensive appearance of the building from afar.

Figure 15: Corner quoins are rock faced below and smooth faced above belt layer. Source: Photographs by author.

Figure 16: Rock face finish (left) and smooth faced finish (right). Source: Photographs by author.

The Towers

There were two four-story stone towers on the opposite corners of the building, with a flat roof covered with copper, a stone parapet and battlements, and a roof hatch. A large metal clock face was added on the southwest tower faces on the fourth level (Figure 17). The access to the tower is through the first floor of the main building. Inside the tower, the space was 14 feet by 14 feet and there were four floors leading to the clock winding mechanism. Access between floors is by wooden ladder-type stairs.[5]

Figure 17: Southwest tower.
Source: Photograph by author.

Originally, the fortress had three stories with a crenelated roof, two towers and two small turrets, one at the southwestern corner and one at the northwest corner. The turrets (Figure 18) were supposedly used as look-out towers.

Figure 18: Northwest turret circa 1910.
Source: Benicia Historical Museum.

The Interior

The lower floor of the building was intended for the storage of gun carriages; the second and third floors were designed for storage of small arms and had glass windows.[6] The second floor was supported by two rows of decorative cast iron columns running lengthwise. These can be seen in the debris from the explosion and fire of 1912 which is also discussed in a later chapter (Figure 19 and 20). The second-story space was completely clear of supports (Figure 21). This was made possible by suspending the entire third-story floor from metal rods hung directly from the roof trusses.[7] There were two sets of stairs between floors.

The electrical power station in Benicia was built in 1883[8] and shortly thereafter the Benicia Arsenal ran power lines to their buildings. The Clock Tower received power sometime after 1891.

Figure 19: Columns in fire debris, 1912.
Source: Benicia Historical Museum.

Figure 20: Clock Tower lower floor, after rebuilding, circa 1948.
Source: Library of Congress, HABS Survey CA-1828 Image 2.

*Figure 21: Clock Tower upper floor after rebuilding, circa 1948.
Source: Library of Congress, HABS Survey CA-1828 Image 4.*

The stone walls in the interior were undressed sandstone and somewhat crudely placed, especially around the windows. The normally tan colored sandstone was oxidized in the fire of 1912 causing them to turn a red hue (Figure 22).

Figure 22: Appearance of interior sandstone walls.
Source: Photograph by author.

The major alteration in this building was the reconstruction necessary after the explosion and fire in the building in 1912. All of the interior was gutted, leaving a pile of twisted iron and charred wood. Interior surfaces of the walls were badly spalled, and on the south side almost the entire top story was destroyed. The building was rebuilt as a two-story structure and the tower on the northeast corner of the building was truncated and a much lower wooden roof added. The turrets at the northwest and southeast corners were removed. The entire interior was rebuilt using concrete columns to support the heavy timber beams and wood floor of the second level.[9]

During the reconstruction, an Otis elevator was installed in 1918 and the elevator remained functional up until 2015, providing 97 years of service.

The Builders of the Clock Tower

Captain Franklin Callender hired Jeremiah Fuss in September 1858 to be the master builder for the new storehouse. Caleb Merrill, who was already employed by the Army, was selected as master stone mason (Figure 23).

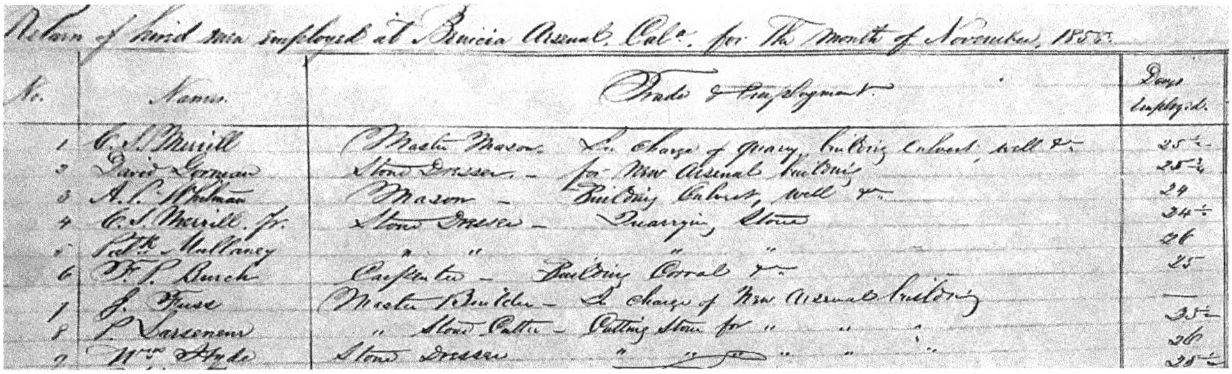

Return of hired men employed at Benicia Arsenal, Cala. for the month of November, 1858.

No.	Names.	Trade & Employment	Days employed.
1	C. S. Merrill	Master Mason – In charge of quarry, building culvert, well &c.	25½
2	David Gorman	Stone Dresser – for New Arsenal building	25¾
3	A. L. Whitman	Mason – Building culvert, well &c.	24
4	C. S. Merrill Jr.	Stone Dresser – Quarrying stone	24½
5	Patk. Mullaney	" " " "	26
6	F. P. Burch	Carpenter – Building corral &c.	25
7	J. Fuss	Master Builder – In charge of New Arsenal building	—
8	P. Larsenen	" Stone Cutter – Cutting stone for " " "	25½
9	Wm. Hyde	Stone Dresser " " " " "	26
			25½

Figure 23: Army ledger from November 1858 .
listing hired workers C. S. Merrill (line 1) and J. Fuss (line 7).
Source: Benicia Historical Museum.

Jeremiah Allen Fuss was born in Franklin County, Pennsylvania, in 1814. He was a carpenter by trade and worked for the Army as master builder at the armory at Harpers Ferry, West Virginia, from 1841 to 1858. He married Naomi Lavinia Webb in July 1841 and they had five children, four sons and one daughter. The government sold, in a private sale in 1852, fifty-two armory dwelling houses which were purchased by workman living in them.[10] Fuss purchased a house and its lot for $400 during this sale and lived with his family there.[11]

Fuss became a superintendent and architect for the Army Ordnance Bureau. In 1846 he supervised the construction of a new brick building, intended for the use of the citizens of Harpers Ferry as a town market. He must have been popular in town as he was elected the first mayor of Harpers Ferry and served for one year in 1851. That summer, he constructed the wooden bridge over the Shenandoah Canal.[12]

In 1858, Fuss came to the Benicia Arsenal to oversee the building of the new storehouse after he was selected by Commander F. D. Callender as the master builder. He worked with Callender to create a drawing based on Callender's vision of the storehouse with the help of drawings of storehouse buildings sent from other military bases. As building began in 1859, he oversaw the construction of the storehouse which took over a year. In 1860, under orders from Julian McAllister, he made four trips to San Francisco to purchase materials for the buildings and to inspect the cast iron columns for the new storehouse.[13] The census in July 1860 showed Fuss living in the Benicia Barracks with many of the other storehouse workers.[14] In April 1859, Fuss was listed on the deed for transfer of land from E. H. von Pfister to St. Paul's Church as a vestrymen along with Julian McAllister. It is possible he had a hand in constructing the first St. Paul's Church building in Benicia.

At the outbreak of the Civil War, Fuss returned to the East Coast and entered into service as a master builder for the Confederate States (CS) Ordnance Bureau in Richmond, Virginia. In 1862, Fuss became superintendent over the construction of the CS Armory at Macon, Georgia. He was also appointed acting master armorer and commissioned a major in the CS Army. After the war, he remained in Macon and supervised the construction of the Bibb County courthouse, Mercer University, and the local Masonic Temple.[15]

Caleb Strong Merrill was the master mason of Powder Magazines 2 and 10. He was born September 25, 1806, in Shelburne Falls, Massachusetts. His father, Thaddeus, was a stone mason and Caleb took up the occupation.[16] He married Maria Childs in June of 1834 and their first son, Caleb Strong, Jr., was born in September of that year. The Merrill family moved to Monmouth, Illinois where Merrill was an architect and builder and had a contractor business. His wife Maria died at the young age of 34 in 1845 so Merrill brought his young family of five back to Shelbourne Falls. In 1852, Merrill and his 17-year-old son, Caleb Jr., traveled to California across the plains and arrived in the fall. Caleb partnered with Pierre "Peter" Larseneur to build the wooden Benicia Barracks.[17] In 1854, they built the Old St. Mary's Cathedral in San Francisco. In 1855, Merrill entered into a contract with the Army to build the first stone powder magazine. He oversaw the quarry work to excavate and remove sandstone blocks and the construction of the magazine. After completion of the magazine, Merrill and Larseneur were hired by the Army as a stone masons to build the second stone magazine in 1857. Merrill continued to work at the Benicia Arsenal along with his son and, in 1859, he was the master mason in building the stone new storehouse. After his work for the Army, Merrill continued the building and construction business in the San Joaquin Valley.

Based on the Arsenal ledger through June 1859, 102 civilian men were hired by the Army to work on building the Clock Tower, 22 of which had previously been involved in the construction of the powder magazine building 10 (Table I).

Table 1: Nationality of Clock Tower Builders.

Nationality	Total Men	Percent
Irish	57	56
American	23	22
British	3	3
Scottish	2	2
Canadian	2	2
French	2	2
Unknown	13	13
Total	102	100

THE CONSTRUCTION OF THE CLOCK TOWER

The Benicia Arsenal ledger of monthly activities provides a detailed account of the personnel and building construction from 1858 to June 1859.[18] Preparation of the new Arsenal storehouse began in July 1858 when demolition of the Quartermasters Residence on the site began. In August, cleaning away the debris on the grounds was completed and quarrying and dressing stone for the building began. Over the following months the workers continued to quarry stone, grading and leveling the storehouse site. Rains began in October and continued over the next several months, but the 47 workers continued to quarry and cut stones and deliver them to the site located about 2,000 feet away.

In February 1859, excavation of the foundation was finished and laying the stones commenced. The work intensified as 55 hired workers were busy in the quarry and foundation site. In March, 1,000 cubic feet of stone was quarried and hauled to the site. 2,750 cubic feet of stone foundation was laid into place. By the end of April, 5,500 cubic feet of dressed stone was laid on the foundation.

By May, the sides of the storehouse building and the towers were built up as far as the double loophole windows. A new shaft was made for the derrick in the quarry. A derrick is a lifting device with a supporting tower and a boom hinged at its base. Rope lines are connected to the top of the mast to allow for lateral and vertical movement of the boom. Derricks were also made for use in constructing the storehouse. The blacksmiths, C. L. Palmer and James A. McDonell, were busy as 300 picks, and 200 stone cutters and chisels were sharpened. In June, 4,502 cubic feet of stone was laid in the storehouse walls. This month, 70 hired workers worked on the storehouse.

The drawings provided details for the cutting of the stones and, the stones were marked to identify their position in the structure (Figure 24 and 25).

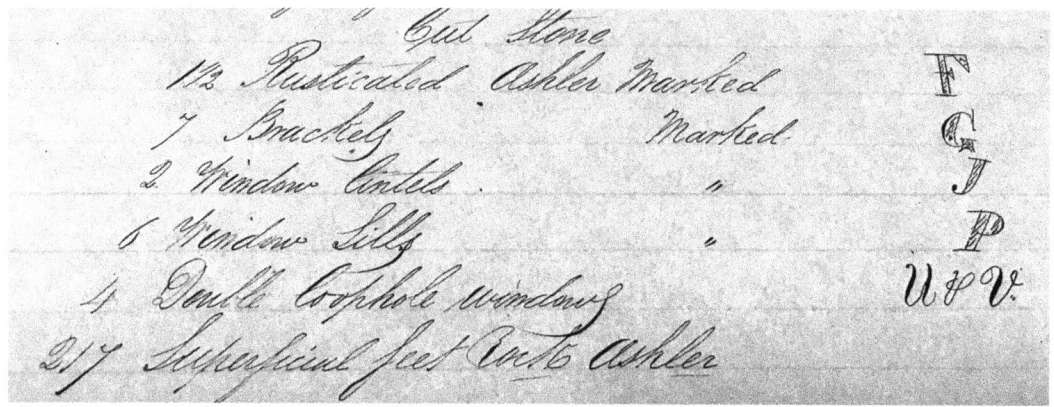

Figure 24: Monthly ledger explaining marking of stones.
Source: Benicia Historical Museum.

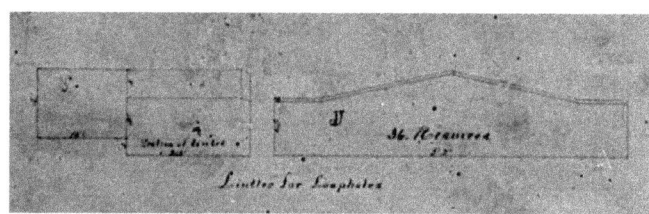

Figure 25: Excerpt from drawing showing details of "J" lintels for loophole windows, "36 Required."
Source: Source: Library of Congress, HABS no. CA-1828-14.

None of these above markings can be seen on dressed sides of the stones. However, one window sill stone on the south side has a marking "5" which is of unknown meaning (Figure 26).

Figure 26: Marking "5" on sill stone.
Source: Photograph by author.

John Bucannon Floyd, Secretary of War, presented a report to Congress for the year ending June 30, 1860, as to the Benicia Arsenal new storehouse status,

The new arsenal storehouse, one hundred and seventy-five feet long by sixty wide, and three stories high, with two towers on the diagonal corners, each eighteen feet square with loop-hole windows for flank defense, built on a plan approved by the Secretary of War, is up to the height of the side walls, the gables and towers remaining yet to be raised to their proper height above that level. The towers, in addition to affording the means of a flank defense of the side and end walls of the building, will also receive the stairways for the different stories of the building, and the rooms in the towers will also afford storage for many small articles and besides, serve as cleaning rooms for small arms, etc. There are loop-hole windows also in the lower story of the building which is intended for the storage of gun carriages; the second and third stories, designed for the storage of small arms, etc,, will have glass. The building is of Benicia sandstone; the exterior face of the wall is rock face, and the interior hammer dressed. The frame for the roof is prepared and ready to be raised; the slates are on hand, and the building will be completed and ready for use in a short time.[19]

The building was completed in 1860. Sitting atop Army Point, the three-story structure was as much a fort as storehouse, with its roof crenelated with battlements and its two towers topped by lookouts. The walls were pierced by apertures for cannons, and by slits for defensive, close-in, musket fire (Figures 27 to 34).

Figure 27: The new storehouse at the Benicia Arsenal, circa 1870s. A horse-drawn wagon is at entrance.
Source: Benicia Historical Museum.

Figure 28: View of two towers, circa 1890.
Source: Benicia Historical Museum.

Figure 29: The storehouse, view from the Commanding Officers Quarters, 1891. Source: Library of Congress, HABS CA-1828-11.

Figure 30: Storehouse view from the Commanding Officers Quarters, 1908. Source: Library of Congress, HABS No. CA-1828-5.

Figure 31: The storehouse, view from the Carquinez Strait, circa 1906.
Source: Benicia Historical Museum.

Figure 32: The storehouse, view of entrance roads, June 1904.
Source: Benicia Historical Museum.

*Figure 33: The storehouse, view of entrance road, circa 1910.
Source: Benicia Historical Museum.*

*Figure 34: The storehouse, circa 1910.
Source: Benicia Historical Museum.*

The Central Pacific Railroad Company completed construction of a rail line route between Benicia and Suisun in December 1879. In April 1885, the railroad was leased to and operated by the Southern Pacific Railroad. This route to Benicia passed below the Clock Tower along Carquinez Strait shore. The locomotive shown in Figures 35 and 36 is a 3001 Class A-1 steam engine, serial number 20795, built by Baldwin Locomotive Company, Philadelphia, Pennsylvania, in August 1902. The locomotive operated on this route before it was retired on December 31, 1924, and scrapped at West Oakland on August 21, 1926.[20]

Figure 35: View of Clock Tower from railroad tracks below circa 1906. Source: Benicia Historical Museum.

*Figure 36: View of Clock Tower from afar with train visible, circa 1906.
Source: Library of Congress, HABS Survey CA-1828. Image 6 by Frank Stumm.*

The Explosion and Fire of 1912

On October 18, 1912, a fire and subsequent explosions starting at 7:30 pm on the second floor totally destroyed the main storehouse at the Benicia Arsenal. The clock captured the beginning of this event as it stopped at 7:30 (Figure 37). For two hours, huge masses of smoke rolled into the sky, gradually giving place to a terrific blaze that shot upwards and could be plainly seen in San Francisco. Every few minutes the neighboring country was shaken by terrific explosions. Despite the numerous explosions, nobody was injured (Figures 37-41).

By 10:30 pm nothing was left of the building but a broken skeleton of walls. The fire consumed 15,000,000 rounds of ball cartridges for small arms and gatling guns, 35,000 army rifles of the latest style, and the complete equipment for thirty thousand soldiers, including uniforms, shoes, blankets, leggings and side arms. The cause of the fire was believed to have been spontaneous combustion in the packing of some of the ammunition. The text of the *San Francisco Call* October 19, 1912 newspaper article[21] is as follows:

STATE ARSENAL
DESTROYED WITH
ARMY SUPPLIES

Old Structure Burns Amid Great
Roar of Exploding Powder
and Shells

{Special Dispatch to The Call}

BENICIA, Oct. 18. — Three black, charred brick walls and a heap of molten brick and ashes is all that remains of the historic old arsenal building at Array point as evidence of a conflagration tonight which is estimated to have caused a loss of nearly $1,000,000 to the United States Army. In short time the ordnance supply station for the Western Division of the army, supply center for the entire Pacific coast and one of the three division arsenals of the country, was completely destroyed.

Flames Leap High
For two and a half hours - 8 p. m. to 10:30 - the flames raged, converting the four story

structure into a roaring furnace that drove back the small force of fire fighters to a distance that made their efforts useless.

The flames surged 100 feet above the structure, which stood on an eminence on the north head of Carquinez Straits. For miles around it shed its gleam. Explosions were numerous and at times like the incessant rattle of musketry fire. Again hundreds of rounds of cartridges were burned in a mighty puff that furnished a spectacular sight.

Toward the end of the conflagration the south wall tottered over, opening a mighty picture of flame to the people of Martinez, the other shore of the straits.

Discovered by Guards

The flames were discovered at 8 o'clock by one of the guards stationed on the grounds. He hurried word to Colonel Golden L. T. Ruggles, chief ordnance officer of the western division and commandant of the arsenal.

With Lieutenant F. H. Miles, the colonel rushed from his house. It was merely a spit of flame from the second story window that the guard had noticed. The two officers and the guard greeted a spectacular scene. As they emerged, the flames had reached the small arms ammunition. A mighty explosion, remarkable in its slight detonation, crashed out the windows in a tremendous puff of powder flame.

Colonel Ruggles, with his force of 45 men, undertook to save some of the equipment. The arsenal was stored with infantry and cavalry equipment, including saddles, blankets, rifles, revolvers, small arms ammunition and the various other items of small equipment. The soldiers attempted to make their way into the structure, but the heat drove them away.

Heat Checks Fight

The fire fighting apparatus at the station was called into play, but the intense heat drove the men back so far that the force of water was not strong enough to reach the structure. The Benicia fire fighting apparatus also was useless.

The fire fighters confined their efforts to extinguishing the grass fires and preventing the flames from spreading to the other buildings.

Nothing was saved from the arsenal. The flames grew so hot that even portions of the brick were molten. Shortly after 9 o'clock the south wall tottered with a heavy crash. This merely increased the draft and the flames leaped higher, making an inspiring spectacle. Half an hour later the flames began to subside and at 11 o'clock there were only smoldering ruins. Colonel Ruggles believes the fire was caused by spontaneous combustion the cartridges stored in the second floor.

The old arsenal was constructed in 1859 and ever since has been used by the government as the main and only arsenal of the west. It was constructed as a three story building, one of the largest brick buildings in the state at the time. A few decades ago a new story was added to the structure to accommodate the increasing demands of the western division of the army.

The arsenal was the supply station for 20,000 men of the western division of the army. Thousands of the old Springfield rifles were stored there, together with a supply of new Springfield rifles large enough for a defending army for the Pacific Coast.

The structure was of brick walls and interior wooden construction of the old type. With the ammunition explosion on the second floor, the flooring gave way. dropping the burning mass down to the first floor where the blankets, saddles and other such equipment is kept. It was this that prevented the entrance of the soldiers within five minutes after the flames were discovered. Had the structure been of concrete type more than half of the contents would have been saved, even after the fire had gained great headway.

Before the arrival of the Benicia fire department several large grass fires were started from the falling sparks. The other structures in the station were threatened, but prompt work on the part of the handful of soldiers stationed there, prevented it.

Several of the soldiers volunteered to fight an entrance into the burning edifice. Seeing that it was an almost unavoidable death for the sake of saving a few pieces of equipment, Colonel Ruggles, after attempting himself to gain entrance, forbade them.

With the first explosion, at the same time the second story floor dropped, the flames surged to the third story and simultaneously to the top floor, where the small arms were kept. The whole structure was in fames within 10 minutes after the first spit of flame had been discovered by the entry.

The powder magazine of the arsenal is stationed half a mile from the main arsenal building. but sparks were carried even this far. A detachment of soldiers was sent to the magazine with wet sacks, and they vigilantly stamped out the little grass fires, with hundreds of tons of high explosive a few feet below.

The three remaining charred walls of the old structure are in such condition that repair will be impossible. The old landmark is practically obliterated. It is expected at the post that the War Department will order a new concrete arsenal for the post.

*Figure 37: The clock damaged by the explosion.
Source: Benicia Historical Museum.*

Figure 38: The storehouse, the morning after the fire.
Source: Benicia Historical Museum.

Figure 39: Internal view of the storehouse damage. Source: Benicia Historical Museum.

Figure 40: The storehouse, after the fire.
Source: Benicia Historical Museum.

Figure 41: Men inspecting the storehouse after the fire.
Source: Benicia Historical Museum.

Rebuilding the Clock Tower

The gutted storehouse was rebuilt as a two-story building and with only one tower. Cleanup and reconstruction took about two years to complete (Figure 42 and 43).

Figure 42: Rebuilding the storehouse after the fire, October 1913.
Source: Benicia Historical Museum.

Figure 43: The rebuilding of the storehouse, November 1913.
Source: Benicia Historical Museum.

After rebuilding, the Army continued to use the storehouse to store munitions, rifles and small arms. The Army referred to the building as the "storehouse with clock."

During WWII, the Clock Tower was extensively used by the Army. In the 1960s, the building was used by the California National Guard, 3632 Ordnance Company (Figure 44). After the City of Benicia acquired the Clock Tower in 1964, the building sat idle into the 1970s (Figure 45).

Figure 44: Back side of the Clock Tower, 1944.
Source: Benicia Historical Museum.

Figure 45: The Clock Tower, 1960.
Source: Benicia Historical Museum.

Figure 46: The Clock Tower, 1970s.
Source: Benicia Historical Museum.

Figure 48: Colonel Julian McAllister.
Source: Benicia Historical Museum.

The Clock

The Seth Thomas clocks, 8 ½ feet in diameter, were installed in two faces of the southwest tower of the storehouse sometime in the 1870s. The clock is visible in the 1878 Atlas sketch of the arsenal (Figure 47). The Seth Thomas Clock Company produced a variety of sophisticated clock styles since 1813 and is recognized as one of the most respected brands of clocks in the world. During its early years, Seth Thomas quickly gained an impeccable reputation for producing masterfully crafted grandfather clocks, and soon after, was also known for designing and producing some of the country's most renowned tower clocks, including one at Grand Central Station in New York.[22]

Figure 47: Atlas view of the arsenal, 1878.
Source: Benicia Historical Museum.

After severe damage to the clocks in the 1912 fire, they were replaced and rededicated to the memory of Colonel Julian McAllister who commanded the Arsenal from 1860 to 1864 and 1867 to 1886 (Figure 48).

An arsenal maintenance log for the storehouse building 29 indicate that a new Seth Thomas clock was installed on November 3, 1936 and put into operation on January 29, 1937 (Figures 49 and 50). After the clock installation, the building was referred to as the "Clock Tower."

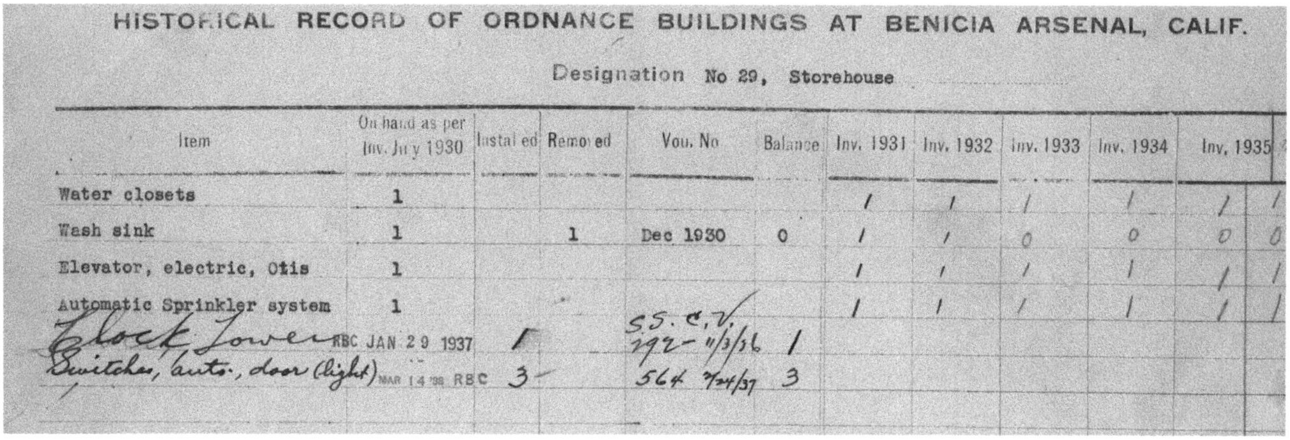

HISTORICAL RECORD OF ORDNANCE BUILDINGS AT BENICIA ARSENAL, CALIF.

Designation No 29, Storehouse

Item	On hand as per Inv. July 1930	Installed	Removed	Vou. No	Balance	Inv. 1931	Inv. 1932	Inv. 1933	Inv. 1934	Inv. 1935	
Water closets	1					1	1	1	1	1	1
Wash sink	1		1	Dec 1930	0	1	1	0	0	0	0
Elevator, electric, Otis	1					1	1	1	1	1	1
Automatic Sprinkler system	1					1	1	1	1	1	1
Clock Tower RBC JAN 29 1937		1		S.S.C.V. 292- 11/3/36	1						
Switches, auto., door (light) MAR 14 '38 RBC		3		564 7/24/37	3						

Figure 49: The maintenance record for storehouse Building 29, 1930s. Source: Benicia Historical Museum.

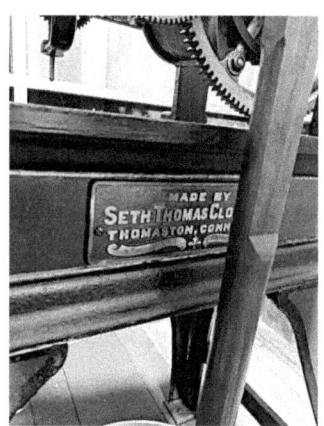

Figure 50: The Seth Thomas clock nameplate. Source: Photograph by Rick Knight.

The clock was operated by a mechanism consisting of a cable, weighted by a cradle with adjustable weights, that slowly unwound on a windlass (Figure 51). The cradle extended from the tower fourth floor down to the second floor (Figure 52). A pendulum was kept in motion by a weighted cannonball and a reference clock was synced to the outside clock hands (Figures 53 and 54). This mechanism had a U-joint that allowed it to operate both clocks in each tower face.

Figure 51: The clock mechanism and windlass.
Source: Photograph by Rick Knight.

Figure 52: The cradle and weights.
Source: Photograph by Rick Knight.

Figure 53: The clock pendulum with cannonball below.
Source: Photograph by Rick Knight.

Figure 54: Reference clock.
Source: Photograph by Rick Knight.

The clock had to be re-wound every six days and was operative as late as June 14, 1967. The clock stopped working and was not restored until 1979. David L Morgan, a member of the National Association of Watch and Clock Collectors of Diablo Valley, returned it to service. Today the clock is still functional, however for access and safety reasons, the clock is no longer wound (Figure 55).

Figure 55: The Clock.
Source: Photograph by author.

The Tunnel

Several sources indicate that there may have been a tunnel that connected the Clock Tower to the Commanding Officer's Quarters (COQ). Ed Hall was interviewed by *The Benicia Herald* in 1994 when he was eighty-eight years old. Hall lived in the Commanding Officer's Quarters from June 1932 until 1935. His stepfather, Colonel Phillip J. R. Kiehl, was the commanding officer of the Arsenal at the time. Hall claimed there was a tunnel between the COQ and the Clock Tower. Hall was told that it was built as an escape route in "old times" in case Indian tribes attacked, but it was never used after 1930.[24]

The National Register of Historic Places #76000534 has the following excerpt about the Commanding Officers Quarters[25]:

A brick tunnel leads from the basement to the near-by storehouse (Clock Tower or Fortress), but there is no record of its having been used as the result of an attack.

The National Register also mentions the tunnel in the Storehouse (Clock Tower Building 29) listing:

The tunnel from the Commandants' house connected to this fortress.

Nineteenth-century photographs of the COQ and Clock Tower show a substantial valley between the two so that a tunnel would have to slope down considerably then slope up to the Clock Tower.[26] The distance of the tunnel would have been 435 feet in length. Army drawings from 1860, however, do not show any evidence of a tunnel in the basement of the COQ.

Benicia city officials have stated that there are no tunnels in the basement or anywhere on the building grounds. However, I have a friend who tells a different story. He lived in Benicia on Jefferson Street from 1986 to 1998. To the east on Jefferson Street there were three deserted old arsenal buildings, known as the Duplex Officers' Quarters, the Captain's Quarters (also known as the Jefferson Street Mansion) and the Commanding Officers' Quarters. At the end of the street was the Clock Tower. He was age 12 at the time and he and his friends would explore these buildings for fun and adventure. He says there were underground tunnels that connected these three buildings. He also says there was a tunnel connecting the COQ and the Clock Tower (Figure 56). This tunnel was about 3 feet wide by 4 feet tall with an arched ceiling made from brick and was damp inside. He and his two friends would wear rain ponchos and use a flashlight to navigate, hunched over, through the dark tunnel (Figure 57). The tunnel leading to the Clock Tower came out in an area near front tower floor, which today can no longer be distinguished. Today, there are no visible traces of tunnels.

Figure 56: Map showing approximate location of tunnel from COQ to Clock Tower, 435 feet long. Source: Diagram by author.

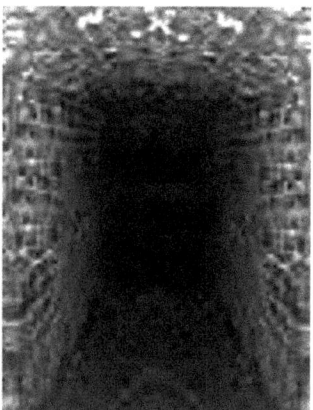

Figure 57: What the tunnel may have looked liked. Source: Depiction by author.

THE CLOCK TOWER TODAY

Figure 58: The Clock Tower, 2021.
Source: Photograph by author.

In 1964, the Clock Tower became the property of the City of Benicia and has since served as a community assembly hall. The facility is used for dances, private parties, receptions, and community functions. The Clock Tower is Benicia's largest facility and has a maximum standing capacity of 601 people and a seating capacity of 536 people.

In 1982 and 1989, seismic upgrades were performed on the Clock Tower. In 2015, the City of Benicia replaced the century-old Otis elevator after irreparable defects were encountered. The new elevator was in compliance with the Americans with Disabilities Act standards.

Appendix A

List of Men Who Built the Clock Tower

The ledger of the Benicia Arsenal monthly activities from January 1858 to June 1859 provides a detailed account of the personnel that the Army hired, their duties, their pay level, and number of days worked.

The worker's trade and pay level was as follows:

- Master builder $7.50 per day
- Master mason: $6.00 per day
- Master stone cutter: $6.00 per day
- Stone cutter: $5.00 per day
- Mason/Stone dresser: $4.00 per day
- Laborer: $2.25 per day

The following is a list of the 102 men who had a part in building the Clock Tower including birthplace and trade:

Name	Birthplace	Trade
Jeremiah Fuss	USA (PA)	Master builder of the Clock Tower
Caleb S. Merrill	USA (MA)	Master mason-in-charge
James Allen	?	Laborer
William S. Birch	USA (MA)	Mason
Adolph Bloudin	?	Stone cutter
Patrick Bradey	?	Laborer
Charles Breen	USA (MA)	Laborer
Richard Brennan	Ireland	Mason
John Breslin	Ireland	Mason ***
Walter Broderick	Ireland	Laborer
F. P. Burch	USA (CT)	Carpenter ***
Peter Burget	?	Laborer
Michael Cain	Ireland	Laborer ***
James G. Caldwell	Ireland	Mason
Thomas Casey	Ireland	Mason
Michael Clarkson	England	Laborer

Robert Clarkson	Ireland	Laborer
Hugh Conley	Ireland	Mason
Cornilius Connoly	Ireland	Laborer
John Conway	Ireland	Laborer
Aeneas Daly	Ireland	Laborer ***
Robert Darling	Scotland	Stone cutter ***
Daniel Day	Ireland	Mason
Patrick Egan	Ireland	Mason ***
John Farrell*	Ireland	Stone cutter ***
John T. Farrell*	Ireland	Laborer ***
Patrick Fegan	Ireland	Laborer
James Fitzharris	Ireland	Laborer
John Flanigan	Ireland	Laborer
Thomas Flynn	Ireland	Stone dresser ***
Michael Gerghegan	?	Laborer
Patrick Gleeson	Ireland	Mason
John Gomo	France	Mason
David Gorman	Ireland	Stone dresser ***
Michael Gorman	Ireland	Laborer
Jacob C. Griffen	?	Mason
John Hicks	USA (NY)	Laborer
Joseph Hilsey	USA (NY)	Mason
James Hogan	Ireland	Laborer
Samuel Hollahan	Ireland	Laborer
John Housley	England	Laborer
James Hughes	Ireland	Laborer
William Hyde	Ireland	Stone dresser
William Johnston	USA (NY)	Laborer
James Keegan	Ireland	Laborer
Christopher Keenen	Ireland	Laborer ***
John Kenney	Ireland	Laborer
Thomas Kinney	Ireland	Laborer
Charles Larseneur	Canada	Stone cutter ***
Peter Larseneur	Canada	Master mason ***
Joshua Leggit	USA (PA)	Laborer
Peter M Livingston	Scotland	Laborer
James Lockwood	England	Laborer
Peter Loughren	Ireland	Laborer

Name	Origin	Occupation
Timothy W. Loughlan	USA (MA)	Laborer
Y H Low	USA (TN)	Laborer
Patrick Henry Lynch	Ireland	Laborer
Patrick Malaney	Ireland	Mason
Phillip Mallett	France	Mason
John Maloney	Ireland	Laborer ***
James Manny	Ireland	Laborer ***
Francis Marshall	USA (TN)	Stone cutter
Hugh McCann	Ireland	Laborer
Denis McCarty	Ireland	Laborer
Peter McCaston	?	Laborer
John McCue	?	Laborer
William McDonald	Ireland	Laborer ***
Jas A. McDonell	Ireland	Blacksmith
Robert McFarland	USA (ME)	Mason
M. M. McGuin	Ireland	Mason
James C. McLaughlin	Ireland	Laborer
Caleb S. Merrill Jr.	USA (MA)	Stone dresser ***
Thomas Moran	USA (IL)	Laborer ***
Patrick Mullaney	Ireland	Mason
William G. Mulschler	?	Laborer
David Murphy**	?	Teamster
David Murphy **	Ireland	Laborer
Eugene Murphy	Ireland	Laborer
Henry Murphy	Ireland	Laborer
John Murray	Ireland	Laborer
Michael Nawman	Ireland	Laborer
William Nicholson	USA (MD)	Laborer
Michael Fitz Patrick	Ireland	Laborer ***
C. L. Palmer	?	Blacksmith
Joseph Petil	Ireland	Laborer
John Potter	USA (NY)	Laborer
John Quinlan	Ireland	Laborer
Edward Quinn	Ireland	Laborer
Daniel Raley	Ireland	Laborer
John Rall	?	Laborer
Michael Ryan	USA (NY)	Stone cutter
F. A. Saddler	USA (NY)	Laborer

James Scanlon	Ireland	Mason
Peter Scanlon	Ireland	Mason ***
Eduard Shanley	?	Stone cutter
James B. Smith	USA (NY)	Laborer
J. C. Stone	USA (NY)	Laborer
Michael Vickers	Ireland	Laborer
John Walsh	Ireland	Laborer
A. P. Whitman	USA (MA)	Mason
David H. Wilson	?	Laborer
John Wyse	Ireland	Laborer

* Two men were named John Farrell, one was a stone cutter and one was a laborer (not related).

** Two men were named David Murphy, one was a teamster and one was a laborer (not related).

*** Worked on Powder Magazine 10

WORKS CITED

1. National Register of Historic Places #76000534, Benicia Arsenal, 1976, p5
2. Historic American Building Survey, CA-1828, Benicia Arsenal Storehouse (Clocktower, Building No. 29), Photographs and Descriptive Data
3. Ibid
4. Ibid
5. Ibid
6. Report of the Secretary of War, John B. Floyd for the year ending June 30, 1860, U.S. Congress, House Documents, 36th Congress, 2nd Session, 1860, Serial Vol. 1079, p. 987.
7. Letter of August 19, 1858, from Commander Franklin D. Callender to Colonel Craig, Benicia Historical Museum
8. Correspondence with Michael Hayes, 2021
9. Historic American Building Survey, CA-1828, Benicia Arsenal Storehouse (Clocktower, Building No. 29), Photographs and Descriptive Data
10. National Register of Historic Places, Harpers Ferry Historic District, Jefferson County, WV, Item 8, page 2
11. Charles W. Snell, *History of the Lower Hall Island and the U.S. Rifle Factory, 1842 to 1885, A Physical History, Volume II*
12. Ibid
13. T*he Charles P. Stone Journal, Ledger of Benicia Arsenal January 13, 1851 through March 25, 1869,* Benicia Historical Museum
14. United States Federal Census, 1860, Benicia, Solano County, California
15. Matthew W. Norman, *Spiller & Burr, The Men*
16. Samuel Merrill, Merrill Memorial, An Account of the Descendants of Nathaniel, an Early Settler of Newbury, Massachusetts, 1938, p451
17. Executive Documents, Second Session of the Thirty-third Congress, 1854-1855, Congressional Edition, Volume 788, *Contracts with the War Department*, page 42
18. Benicia Arsenal Ledger, June 1856 to June 1859, Benicia Historical Museum, catalog 1994.007.0002
19. Report of the Secretary of War, John B. Floyd for the year ending June 30, 1860, U.S. Congress, House Documents, 36th Congress, 2nd Session, 1860, Serial Vol. 1079, p. 987.
20. Don Ross Group website, donrossgroup.net
21. S*an Francisco Call*, October 19, 1912
22. www.SethThomas.com
23. Historic American Building Survey, CA-1828, Benicia Arsenal Storehouse (Clocktower, Building No. 29), Photographs and Descriptive Data
24. James Lessenger, *Commanding Officer's Quarters of the Benicia Arsenal*, 2010, James Stevenson Publisher, p61
25. National Register of Historic Places #76000534, Benicia Arsenal, 1976, p4 & 5
26. James Lessenger, *Commanding Officer's Quarters of the Benicia Arsenal*, 2010, James Stevenson Publisher, p106

About the Author

Allan Gandy is a graduate of California Polytechnic State University, San Luis Obispo, with a degree in Metallurgical Engineering. He has been a resident of Benicia, California for over 40 years and has an interest in its history. Allan is a docent and research associate at the Benicia Historical Museum and the author of a previously published work titled, *The Sandstone Magazine at the Benicia Arsenal, Benicia's Little-known Gem,* (2021 by ALIVE Book Publishing).

Allan is an amateur geologist and has studied the geology of California and volcanology. Another interest of his is the California Gold Rush. He has prospected for gold and taught school children to pan for gold. He has lived in California most of his life and loves the beauty of the California coast, the redwood forests, the Shasta and Lassen volcanic landscapes, the Sierras, and other historical areas throughout the state.

ABOOKS

ALIVE Book Publishing and ALIVE Publishing Group
are imprints of Advanced Publishing LLC,
3200 A Danville Blvd., Suite 204, Alamo, California 94507

Telephone: 925.837.7303
alivebookpublishing.com

www.ingramcontent.com/pod-product-compliance
Lightning Source LLC
Chambersburg PA
CBHW060942170426
43196CB00022B/2962